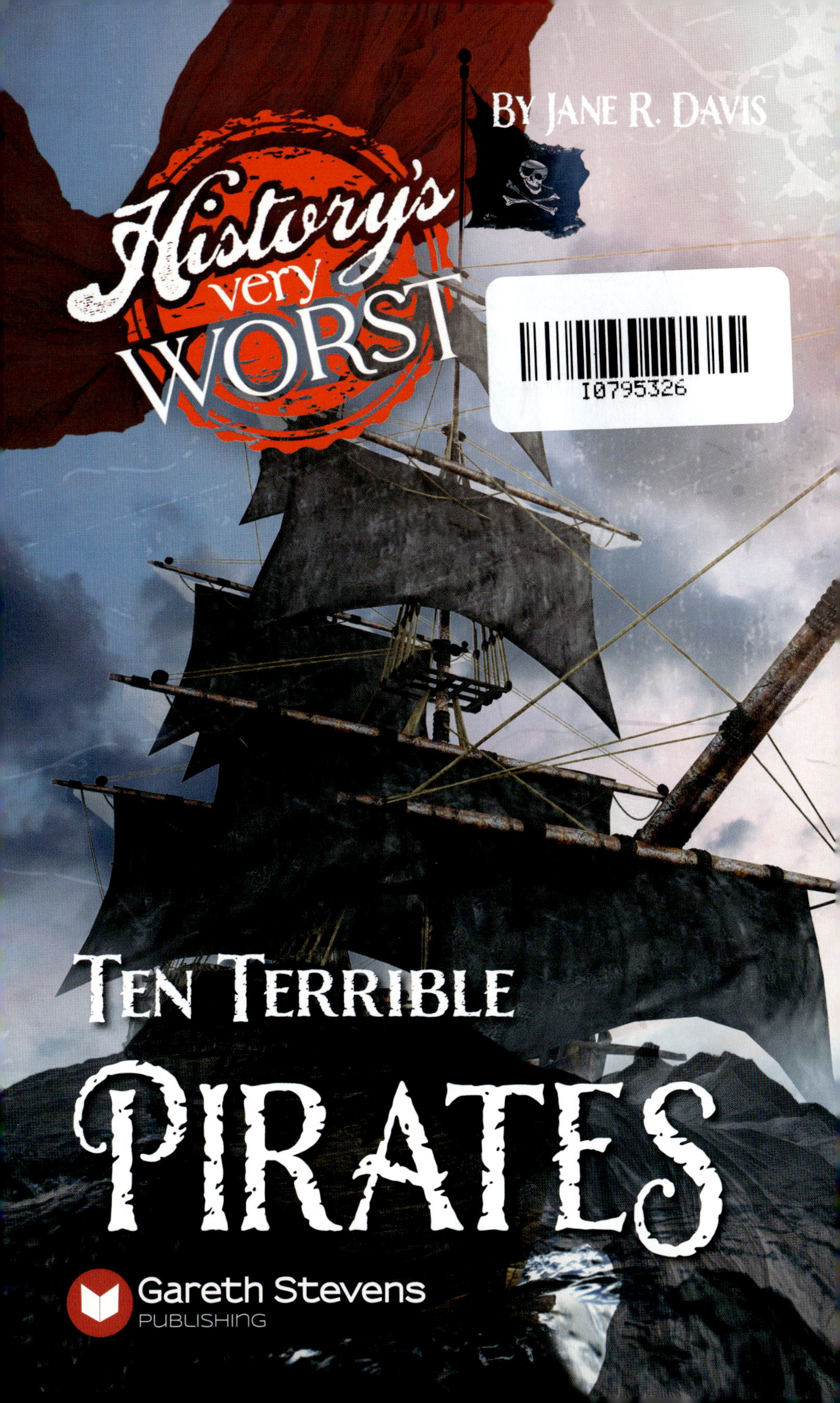

History's very WORST

By Jane R. Davis

Ten Terrible Pirates

Gareth Stevens Publishing

Please visit our website, www.garethstevens.com. For a free color catalog of all our high-quality books, call toll free 1-800-542-2595 or fax 1-877-542-2596.

Library of Congress Cataloging-in-Publication Data
Names: Davis, Jane R., author.
Title: Ten terrible pirates / Jane R. Davis.
Description: Buffalo, New York : Gareth Stevens Publishing, 2026. | Series: History's very worst | Includes index. | Audience: Grades 4-6
Identifiers: LCCN 2025011700 (print) | LCCN 2025011701 (ebook) | ISBN 9781482472714 (library binding) | ISBN 9781482472707 (paperback) | ISBN 9781482472721 (ebook)
Subjects: LCSH: Pirates–Biography–Juvenile literature. | Pirates–History–Juvenile literature. | LCGFT: Biographies
Classification: LCC G535 .D325 2026 (print) | LCC G535 (ebook) | DDC 910.4/5–dc23
LC record available at https://lccn.loc.gov/2025011700
LC ebook record available at https://lccn.loc.gov/2025011701

Published in 2026 by
Gareth Stevens Publishing
2544 Clinton Street
Buffalo, NY 14224

Designer: Tanya Dellaccio Keeney
Editor: Kristen Nelson

Photo credits: Cover Denis—S/Shutterstock.com; p. 4 https://upload.wikimedia.org/wikipedia/commons/e/eb/Åland_piratflagga_(color_fix).png; p. 5 https://upload.wikimedia.org/wikipedia/commons/7/77/General_History_of_the_Pyrates_-_Captain_Bartholomew_Roberts_with_two_Ships.jpg; p. 7 https://upload.wikimedia.org/wikipedia/commons/b/b1/A_French_Ship_and_Barbary_Pirates_%28c_1615%29_by_Aert_Anthoniszoon.jpg; pp. 9, 25 German Vizulis/Shutterstock.com; pp. 10, 15 Shutterstock AI Generator/Shutterstock.com; p. 11 PorcupenWorks/Shutterstock.com; p. 13 (bottom) Peter Hermes Furian/Shutterstock.com; p. 13 (top) slavantonov/Shutterstock.com; p. 17 https://upload.wikimedia.org/wikipedia/commons/1/12/Capture-of-Blackbeard.jpg; p. 19 https://upload.wikimedia.org/wikipedia/commons/0/08/Charles_Vane%2C_Defying_the_Governor%2C_from_the_Pirates_of_the_Spanish_Main_series_%28N19%29_for_Allen_%26_Ginter_Cigarettes_MET_DP835025.jpg; p. 20 Andrejs Marcenko/Shutterstock.com; p. 21 https://upload.wikimedia.org/wikipedia/commons/b/b3/Rackham%2C_Jack.jpg; pp. 22, 23 Yogi Black/Alamy Images; p. 27 IanDagnall Computing/Alamy Images; p. 29 (top) Blue Sky Imagery/Shutterstock.com; p. 29 (bottom) sabs.e/Shutterstock.com.

Some of the images in this book illustrate individuals who are models. The depictions do not imply actual situations or events.

Printed in the United States of America

CPSIA compliance information: Batch #CSGS26 For further information contact Gareth Stevens, at 1-800-542-2595.

CONTENTS

Words in the glossary appear in **bold** the first time they are used in the text.

AHOY, MATEYS!

When you imagine a pirate, you might picture a man with a three-cornered hat, an eyepatch, and a parrot on his shoulder. Maybe he's standing on the deck of a ship flying a black flag with a skull and crossbones on it—the Jolly Roger.

Accounts in ***A General History of the Pyrates*** by Charles Johnson and newspapers reported thrilling stories of the pirates of the early 1700s. However, many of these stories were made to seem more exciting than they actually were.

Your ideas probably come from movies and stories about pirates! In fact, pirates were men *and* women who robbed, or raided, ships carrying goods and treasures. Many **terrorized** the seas from about 1650 to about 1730 around the **Caribbean**. Some call this "golden age of piracy," though piracy occurred before and after that time.

Pirates in stories and movies seem like they only head to land to sell their treasure and cause trouble. Real pirates had families, land, and sometimes businesses to run.

Some men who turned to piracy worked on ships, such as with the British Royal Navy. Low-ranking sailors were often treated poorly. They might **mutiny** and then decide to become pirates to make more money. Other pirates may have been looking for an adventure.

The pirates still remembered today are those who made their fortune on the high seas or otherwise made a splash. Many of them did terrible deeds!

THE UGLY TRUTH

MOST PIRATES DIDN'T OFTEN HAVE THE CHANCE TO ATTACK SHIPS THAT CARRIED "TREASURE," SUCH AS SILVER. THEY MOSTLY WENT AFTER SHIPS THAT HAD EVERYDAY ITEMS SUCH AS CLOTH AND WOOD!

Pirates stole goods from ships and sold them to nearby colonies and islands. Sometimes they stole and sold enslaved people too.

FRANCIS DRAKE

Pirate or privateer? Sir Francis Drake had permission from the English queen to raid Spanish ships. This made him a privateer, or someone hired to upset an enemy's trade while at war. To the Spanish, Drake was a **ruthless** pirate attacking their ships and stealing their goods.

Drake had a history of piracy before he worked for the English crown. During the 1560s he raided ships near France and Africa. Later, in 1577, he led an expedition, or journey, around the tip of South America for England. He wanted to explore—but also continue his piracy against the Spanish!

THE UGLY TRUTH

DRAKE HELPED GET ENGLAND STARTED IN THE TRADE OF ENSLAVED PEOPLE. HE MAY HAVE ENSLAVED MORE THAN 1,000 AFRICANS.

On his way around South America, Drake attacked Spanish settlements and ships, stealing silver, gold, pearls, and gems. He continued his journey and ended up sailing around the globe!

GRACE O'MALLEY

Grace O'Malley is sometimes called the "pirate queen of Ireland." She lived during the mid-1500s in western Ireland. Her family already practiced piracy! O'Malley stayed in the business partly to protect her family's land in Ireland. O'Malley and her crews made people pay to pass through "her" waters. This toll would guarantee, or promise, a ship's safety.

THE UGLY TRUTH

WITH INCOME FROM BOTH PIRACY AND TRADE, GRACE O'MALLEY MADE A LOT OF MONEY!

Grace O'Malley would have known the seas and weather around Ireland well. This made her good at controlling her territory!

O'Malley also raided cattle farms along Ireland's coast as well as raided the islands of Scotland. She commanded huge ships that could hold 200 crew members. They pirated in Spain too, but O'Malley also took part in trade there.

HENRY MORGAN

Henry Morgan reached the Caribbean around 1655. He sailed for the English as a privateer in that area. He became a commander of his own ships.

Morgan and his crews began to raid Spanish settlements, including in Mexico and Cuba. They found gold, silver, and other treasures. In addition, Morgan stole enslaved people and sold them.

In 1671, Morgan and his crew took one of the biggest Spanish cities in the Americas, Panamá. Morgan's crew stole from the city and then burned it down. However, Spain and England had just agreed to peace. Morgan was now seen as a pirate—and he was arrested for it in 1672.

After the raid of Panamá, Morgan left his men and took what they had stolen for himself!

THE UGLY TRUTH

In 1674, King Charles II of England needed Morgan's help in the Caribbean. So, Morgan was **knighted** and named lieutenant governor of Jamaica!

WILLIAM KIDD

The British government hired William Kidd as a privateer in 1689 to arrest pirates! He went on to oppose pirates—or enemy privateers—near Massachusetts and New York too. In 1695, Kidd headed to the Indian Ocean to stop pirates there.

Kidd seems to have turned to piracy himself a couple years later. He may have seen the unprotected ships carrying costly goods in the Indian Ocean and wanted to find his fortune. In 1698, Kidd captured a valuable ship called *Quedagh Merchant* and sailed it to the Caribbean. Not long after, he learned he had been named a pirate and was arrested!

There's a story that William Kidd, pirate hunter likely turned pirate, may have hidden some treasure! **Archaeologists** thought they may have found some of it in 2015.

THE UGLY TRUTH

WILLIAM KIDD WAS KILLED FOR THE CRIME OF PIRACY IN 1701, BUT IT'S UNCLEAR TO TODAY'S HISTORIANS WHETHER HE WAS GUILTY.

BLACKBEARD

Blackbeard is perhaps the most famous pirate remembered today. He worked the waters near the colonies of Virginia and the Carolinas as well as the Caribbean in the early 1700s. His real name may have been Edward Teach (or Thatch), but pirates often used fake names. Blackbeard was named for his long black beard!

Around 1716, Blackbeard turned a captured ship into *Queen Anne's Revenge*. This was his main ship, but he commanded others too. He and his crew raided ships and nearby islands for treasures. Around 1718, he also blocked the port of Charleston, South Carolina, and took **hostages**. He traded them for medicine!

Did Blackbeard leave behind buried treasure? Stories say so—but historians don't think that's true.

Blackbeard may have started out as a privateer before he became a pirate. He died fighting members of the British Navy in 1718.

CHARLES VANE

Charles Vane struck fear into the hearts of many—including other pirates. He started his career as a pirate around 1716 on the ship of another pirate, Henry Jennings. Vane went on to captain his own vessels and terrorize the Caribbean and eastern coast of North America.

Vane escaped capture many times as he **seized** ships and gathered wealth. He became known for **torturing** the people who sailed on the ships he captured. Vane didn't treat his own crew well either. They mutinied and left him on an island. When he was picked up by a ship, Vane was turned over to the **authorities**.

Vane made a fortune as a pirate, likely stealing more than $2.3 million in today's money.

THE UGLY TRUTH

AFTER A TRIAL IN PORT ROYAL, JAMAICA, VANE WAS FOUND GUILTY OF PIRACY. HE WAS HANGED IN MARCH 1721. HIS REMAINS HUNG IN A CAGE TO WARN OTHERS AGAINST THE CRIME OF PIRACY.

CALICO JACK RACKHAM

John "Calico Jack" Rackham has a bigger-than-life **reputation**. In truth, he wasn't the most successful pirate of the "golden age"! He didn't grow rich from piracy, and he wasn't terribly feared by other pirates.

THE UGLY TRUTH

RACKHAM HAD BEEN PARDONED, OR FORGIVEN HIS CRIMES, AROUND 1719. HE TRIED TO LIVE A NORMAL LIFE IN THE BAHAMAS—BUT ENDED UP GOING BACK TO PIRACY!

Calico Jack Rackham is in *Pirates of the Caribbean: The Curse of the Black Pearl*—well, sort of. In the first scene of the movie, a person's remains hang on the pirate ship the *Black Pearl*. Those remains are supposed to be Calico Jack!

While pirating in the Caribbean in the early 1700s, Rackham gained two claims to fame that have made him a pirate legend. First, he made use of the Jolly Roger popular. Second, he had two crew members that were women! One of them, Anne Bonny, fell in love with Rackham. The other, Mary Read, fooled Rackham by dressing like a man.

ANNE BONNY

Anne Bonny was born in Ireland around 1698. She moved to South Carolina, then a British colony, as a child. Even then she's said to have had a **violent** nature. In 1718, she married James Bonny. They headed to the Bahamas where he worked to find and turn in pirates to the authorities.

Anne Bonny went back to South Carolina after her dangerous life on the high seas. She lived until age 84!

THE UGLY TRUTH

RACKHAM, BONNY, AND READ WERE CAPTURED IN 1720. THEY WERE ALL FOUND GUILTY OF PIRACY. RACKHAM WAS HANGED. BONNY AND READ WEREN'T PUT TO DEATH BECAUSE THEY WERE BOTH PREGNANT.

Soon after, Anne Bonny met John Rackham and joined his crew of pirates. She dressed in men's clothes when they attacked other ships. She befriended Mary Read, and the two fought fiercely together, each holding a gun and a **machete**.

EDWARD LOW

Edward "Ned" Low is seen as one of the most terrible pirates to ever sail the high seas. Not only that, unlike many other pirates, he was never caught by the authorities!

Low sailed along the Atlantic coast of North America and in the Caribbean, as well as Portugal, during the 1720s. His piracy began after he and other crew members mutinied while working on a British ship. He quickly found success—and earned a bad reputation.

When Low seized ships, he would torture the crews or simply kill them. He was said to cut off the ears and noses of people he captured.

THE UGLY TRUTH

LOW MAY HAVE FLOWN A SLIGHTLY DIFFERENT JOLLY ROGER. IT'S SAID HIS HAD A RED SKELETON ON IT!

Low treated his crew badly. They mutinied and left him on an island! No one knows for sure what happened to him after that.

Cheng I Sao

Piracy happened beyond the Atlantic, and long after the 1700s. Cheng I Sao was born in South China in the late 1770s. She married a powerful pirate. Upon his death in 1807, Cheng I Sao took over the ships and thousands of pirates they had commanded together.

Cheng I Sao created a code for her pirates to follow. It had **harsh** punishments. Pirates who didn't obey orders were beheaded. Those who skipped work and didn't tell their commander would have an ear cut off! This code kept Cheng I Sao's pirates in line as they grew their fortune and fought the Chinese Navy and others.

Cheng I Sao's name is sometimes written as Shih Yang, Mistress Ching, and in other ways.

THE UGLY TRUTH

CHENG I SAO AND HER CREWS WERE PARDONED IN 1810. SHE WENT ON TO LIVE TO AGE 69!

THE END OF PIRACY?

The "golden age of piracy" ended in the mid-1700s partly because colonies in the Americas were growing. Governments paid closer attention to what was happening at their ports. Countries around the world worked harder to arrest pirates. Pirates' terror was hurting trade.

Some pirates were put to death for their crimes. Others were pardoned. Governments hoped pirates would go free—and stop pirating! Some did. But, other times pirates just moved on to parts of the world where no one was trying to stop them!

Piracy still happens today, though much less. In 2024, more than 110 ships were attacked by modern pirates!

The history of pirates still interests people today.

GLOSSARY

archaeologist: A scientist who studies past human life and activities.

authorities: The government bodies and people in charge of policing.

Caribbean: Having to do with the many islands found in the Caribbean Sea.

harsh: Overly severe.

hostage: Someone who is being held against their will.

knight: To make a knight, or

machete: A large, heavy knife.

mutiny: To rise up against a leader.

reputation: The views that are held about something or someone.

ruthless: Cruel, or having no mercy.

seize: To take by force.

terrorize: To use violence and fear to gain something.

torture: To cause someone pain to force them to do or say something.

violent: Having to do with the use of force to harm someone.

FOR MORE INFORMATION

BOOKS

Gunderson, Jessica. *Blackbeard, Feared Pirate Captain*. North Mankato, MN: Capstone Press, 2025.

Kim, Carol. *Pirates: Spot the Myths*. North Mankato, MN: Capstone Press, 2024.

Lewis, Leigh. *Pirate Queens: Dauntless Women Who Dared to Rule the High Seas.* Washington, DC: National Geographic, 2022.

WEBSITES

Blackbeard

https://www.rmg.co.uk/stories/topics/blackbeard-edward-teach-pirate

Find out more about this fearsome pirate.

Francis Drake

https://exploration.marinersmuseum.org/subject/francis-drake/

Read even more about this pirate from history.

INDEX